Rabbits

by Mervin F. Roberts
**World renowned author of books on rab-
bits and turtles and other pets.**

Acknowledgments

The author is indebted to Ann M. Johnson and to Jason Tolderlund for their gracious help.

Distributed in the UNITED STATES by T.F.H. Publications, Inc., 211 West Sylvania Avenue, Neptune City, NJ 07753; in CANADA by H & L Pet Supplies Inc., 27 Kingston Crescent, Kitchener, Ontario N2B 2T6; Rolf C. Hagen Ltd., 3225 Sartelon Street, Montreal 382 Quebec; in ENGLAND by T.F.H. Publications Limited, 4 Kier Park, Ascot, Berkshire SL5 7DS; in AUSTRALIA AND THE SOUTH PACIFIC by T.F.H. (Australia) Pty. Ltd., Box 149, Brookvale 2100 N.S.W., Australia; in NEW ZEALAND by Ross Haines & Son, Ltd., 18 Monmouth Street, Grey Lynn, Auckland 2 New Zealand; in SINGAPORE AND MALAYSIA by MPH Distributors (S) Pte., Ltd., 601 Sims Drive, # 03/07/21, Singapore 1438; in the PHILIPPINES by Bio-Research, 5 Lippay Street, San Lorenzo Village, Makati Rizal; in SOUTH AFRICA by Multipet Pty. Ltd., 30 Turners Avenue, Durban 4001. Published by T.F.H. Publications Inc., Ltd. the British Crown Colony of Hong Kong.

CONTENTS

CREDITS
All photographs by Michael Gilroy unless otherwise indicated. Material has been adapted from the Carnation Milling Co. book "Rabbits . . . A Carnation Plan for Raising Rabbits"; the Glick Manufacturning Company; and the U.S. Department of Agriculture Handbook No. 490.

Many different animals are used for laboratory and experimental work. Monkeys and rabbits are just two examples. Fortunately, after the experiments are completed many of the animals are sold as pets. This is especially true of rabbits that are used for dietary and toxicity tests.

People and Rabbits

It is difficult for a single human to be objective about an individual rabbit. It is still more difficult for randomly selected humans to agree about what rabbits mean to all of them. Some Chinese revere rabbits, especially during their Festival of the Moon. American Algonquin Indians believe that the Great Hare, Michabo, created the earth, water and fish. Some of these Indians believe, also, that Michabo re-established the earth after The Flood. European mythology gives us the Easter Bunny and has it laying colored eggs. Many people in Australia and New Zealand consider the rabbit as they would a terrible plague. Then there are people who selectively breed rabbits in order to exhibit them in shows for color or pattern or size or shape or how their ears flop or don't flop. Also there are people who breed rabbits for their meat or their wool or their skins or their serums and hormones and antitoxins in order to determine pregnancy or fight disease or study genetics. And don't forget the Easter Bunny pet shop trade, consisting of the people who buy them and the people who end up with them.

Let's take a look at rabbits now from a scientific point of view by first considering how they resemble and how they differ from other animals.

What could be cuter than walking your pet rabbit down the street like a dog? Your petshop even sells leads for rabbits; they are the same leads designed for dogs.

People and Rabbits

The Tree of Life

Kingdom: **Animalia**—This is the kingdom that excludes plants, fungi and some microscopic organisms that are both plant-like and animal-like.

Phylum: **Chordata**—This is the phylum of animals with spinal cords. Say goodbye here to worms, mollusks, sponges, jellyfishes, insects and shellfishes.

Sub-phylum: **Vertebrata**—A sub-phylum of chordates with backbones. Here exits the sea squirt.

Class: **Mammalia**—The class of vertebrates that nurse their young. Here we eliminate the birds and fishes and reptiles and amphibians.

Subclass: **Eutheria**—A subclass of mammals with placentas. Leave out marsupials and monotremes.

Order: **Lagomorpha**—An order of placental mammals with doubled incisor teeth in the upper jaw.

Family: **Leporidae**—The family of rabbits and hares but excluding picas.

Genus: **Oryctolagus**—A genus in the large family Leporidae. There are still other genera of hares and rabbits which are distantly related but are located elsewhere for technical reasons.

Species: **cuniculus**—The species described by Linne (also known as Linnaeus); it encompasses all our domestic varieties of rabbits.

So here is the rabbit *Oryctolagus cuniculus*.

The generic name is derived from the Greek and means "burrowing hare." The specific name is from the Latin and has two translations. You can read it as "rabbit" or as "cave or subterranean passage." So thus we end up with a rabbit or hare whose home is underground.

Going back to the lagomorphs for a moment, you might have wondered why the teeth were mentioned. Well, classifiers of mammals find that they can make some order out of nature by tabulating the number and arrangement of teeth.

The interesting thing about a lagomorph is that there are two incisors on each side of the upper jaw but only one in the lower. Many people never see that second top incisor, because it is not where one would expect to find it. The second upper incisor of a rabbit is behind, not next to, the first. This is an important structural detail that we use to separate lagomorphs from rodents. For 100 or more years, rabbits were included in the order of rodents (Rodentia), but now the best opinion sets them apart.

Like rodents, rabbits have incisors

People and Rabbits

that grow constantly, but these front teeth are worn down by the foods they chew. If a rodent or a lagomorph were deprived of something to gnaw, its incisors would grow until they overlapped; and then eating would be impossible. Such an animal would starve to death.

Wild Rabbits

From the tree of life you could see that there are many lagomorphs with long ears and short tails, but there is only one species which has been domesticated; this is *O. cuniculus*. It does not readily hybridize with other species. It was originally native to the lands bordering the western Mediterranean and has been domesticated for at least 2000 years. Roman armies carried rabbits to France. Today, the British rabbit described by Richard Adams in that marvelous novel *Watership Down* is a direct descendant of animals carried to the British Isles either by the Romans or during the Norman conquest. There are a few other introduced populations that took hold and did nicely. Australia has had its share of rabbits, and so has New Zealand; these rabbits were known to have been introduced over 100 years ago. This Australian species is identical to the wild rabbit of the western Mediterranean and to the wild rabbits introduced to France and the Low Countries by the Romans and to Great Britain by either the Romans or the Normans. In the U.S. there is a population of feral European rabbits on the island of San Juan, Washington, and for

A jack rabbit.

7

People and Rabbits

some years there was and may still be a small colony on Falkner, an island in Long Island Sound off the Connecticut shore. Through the years, people have tried to play God with rabbits by planting them where there was thought to be a need. Where the rabbits thrived, they often wiped out their food supply or caused other catastrophic damages to their environment.

One variety of this domesticated species is called the Belgian Hare, but the name has no zoological significance. It is still *O. cuniculus*.

I might mention in passing that while *O. cuniculus* produces blind and naked young, some other lagomorphs deliver fully furred pups which get about quite nicely when they are only a few days old.

If you ever encounter a wild cottontail or hare or coney or rabbit or any other long-eared buck-toothed jumping animal, leave it where you found it. These wild distant relatives of the domesticated *O. cuniculus* are worse than useless in captivity. Most will not survive more than a few days, and many will transmit diseases that could kill you or your caged stock. The other side of the coin is that every domesticated rabbit variety or type or style or race or size is of that one same species, just as all domesticated dogs or cats or chickens are all of one species.

Rabbits in Captivity

Rabbits do wonderfully in captivity; we would be well advised to keep it so. How well do they do? Well, look at the cost of their meat —it is hardly more per pound than chicken. This is amazing, since the doe rabbit produces no more than 50 youngsters per year against about 200 or 300 eggs per year from a good Leghorn hen.

Even though the chicken egg can produce a broiler or a fryer with no parental help, the efficiency of the rabbit is said to be better than that of a chicken when it comes to making meat from plants. A pound of rabbit meat can be generated from less than three pounds of dry rabbit pellets. The best chicken can do no better on grain and grain costs more than pellets. Add to the value of the rabbit's meat what can be gotten for the pelt and the manure; consider that they don't crow or cackle, and you will have discovered economics that were known to the Romans under the Emperor Hadrian, nineteen centuries ago.

Selecting Rabbits

Decide what you want from your rabbit—a domestic pet, a producer of wool, a wholesome food, a challenge for selective breeding to be exhibited in competition, a pelt. Some varieties of rabbits are multiple purpose and others are especially valuable for just one reason. Regardless, you should be discriminating and ruthless when you select what you must pay for and cage and feed. Buy the best you can afford—the cheapest and fastest and easiest way to improve your stock is by buying the best you can when you first get started.

Visit a knowledgeable pet dealer or go to a sanctioned rabbit show and see what wins prizes. Look at what breeders have to offer. Don't buy the first animal that comes along until you have studied something to compare with it.

Look for an animal with a smooth body, no lumps, bumps, wounds, sores or bare spots. The eyes should be bright and not weepy. The nose should be clean. The teeth should be even and unbroken. Count the toes and toenails. Look at the vent; it should be clean and dry. The body should be fleshy but not necessarily fat. The animal should look bright and alert. If it is not, don't buy it.

If an experienced rabbit keeper reads this he might take issue with the phrase "fleshy but not fat."

Your local pet dealer can obtain any kind of rabbit you want (within reason) including, left to right: long-eared Lops, two-colored Dutch rabbits in various colors plus white, and long-haired English or French Angoras.

Selecting Rabbits

There are some breeds of rabbit like the Belgian Hare that have a real greyhound appearance—an extra ounce of meat anywhere would seem out of place. Others like the commercial meat-producing Californians and New Zealands must be solid and fleshy. These animals will weigh over four pounds when they are only eight weeks old.

Read rabbit books. Look at the pictures. Visit breeders. Join an association or breeders club. Go to a rabbit show. Ask a knowledgeable dealer. Purchase rabbits only after you know what you are looking at. Eat your mistakes; don't breed from them.

Breeds

Obviously, an author of a rabbit book hopes you will read it before you acquire any animals but for most of us, that's not the way it happens. If you started with an animal and then set about to read about it, all is not lost. Most rabbit keepers have more than one breed and many switch once or twice as they get started. One nice thing about rabbits is that with very few exceptions, they all get the same treatment and they all respond in about the same manner.

Here are all the A.R.B.A.-recognized breeds including the colors (usually called varieties) that are available. When you finally settle down with a breed, you owe it to yourself and to your animals to join a specialty club and learn precisely what to aim at.

American—The ideal buck weighs nine pounds, the doe ten. This breed originated in the U.S. largely through the efforts of L.H. Salisbury of Pasadena, California, around the end of the First World War. Its main claim to fame is its solid, dark, clear slate-blue color. A white variety also exists.

Angora, English—The ideal buck weighs six pounds, the doe seven. Angoras are long-haired rabbits. There are no other long-hairs but Angoras. They probably originated in Ankara, Turkey, and have been known at least since 1765. Angoras produce "wool;" all other breeds produce "fur." It is for this wool, and for show, that Angoras are raised. This English breed has a silky wool; by contrast the French Angora has a coarser coat. The recognized colors are White, Black, Blue and Fawn.

Angora, French—The ideal bucks and does weigh eight pounds or more. The wool is coarser than that of the English Angora and the

Selecting Rabbits

preferred length is two and three-quarters inches. The color varieties are White, Black, Blue and Fawn.

Belgian Hare—The ideal weight for both sexes is eight pounds. This animal is exactly the same species as all the other domestic rabbits; it is only the popular name that is confusing. This was an especially popular breed in the U.S.A. and in Great Britain 75 years ago although it originated in Flanders, a section of Belgium. The color is a rich red with a deep tan or chestnut shade. The fur is coarse and short.

Bevern—The ideal weight for bucks is nine pounds and does weigh one pound more. This breed originated in the town of its name near Antwerp, Belgium. Whites have blue eyes. The other colors are a clear blue and glossy jet-black.

Californian—The ideal weight of a buck is nine pounds and does ideally weigh nine and a half. This very popular meat, fur and show breed was given a working standard in 1939. The breed is always white with black ears, nose, feet and tail.

Champagne D'Argent—Formerly known as "French Silver," this is an old recognized breed. Bucks ideally weigh ten pounds and does weigh

---

Dutch rabbits weight about 4½ pounds. They are found in color combinations of white with black, blue, chocolate, tortoise, steel gray and gray.

11

ten and a half. The young are black and as they mature they develop their typically silver color. They should look silvery, never yellow or brassy. They are raised for both show and meat.

Creme D'Argent—The ideal buck weighs nine pounds, the doe a pound heavier. This is a breed of French origin. This is an orange-colored rabbit with the silvery appearance of the Champagne D'Argent.

Checkered Giant, American—The ideal buck weighs eleven pounds or more and the doe weighs at least twelve. This is a really big rabbit. It resembles the English, but it is somewhat less blocky. The two recognized colors on the white body are either black or blue. The precise location of the checkering is important for show animals. Checkereds are also raised for their fur.

Chinchilla—The ideal weight for a standard buck is six and a half pounds. A standard doe should weigh seven pounds. There are three varieties of this breed: Standard, American and Giant. Chinchilla rabbits came to the U.S.A. in 1919 and have been popular ever since then. They are true rabbits. The name comes from the fur which resembles that of the South American animal which is not a rabbit. The real Chinchilla fur is a dark slate-blue undercolor with a pearly intermediate and a black band and a very light band. Chinchilla rabbits are raised for show and for fur. The American variety is larger than the Standard; bucks are ideally ten pounds and does are eleven. The Giant variety of Chinchilla rabbit is still larger; ideal bucks weigh thirteen or fourteen pounds and does a pound more.

Dutch—Bucks and does ideally weigh four and a half pounds. The Dutch pattern found on many animals is desired and the uncolored part is always white. The recognized varieties are Black, Blue, Chocolate, Tortoise, Steel Gray and Gray. It is said to have originated in Holland, but evidence is lacking. Dutch rabbits are raised for showing and for use in laboratories.

English Spot—The ideal buck weighs six pounds and the doe seven. There are tiny spots in a chain pattern on a white background. The recognized show varieties are Black, Blue, Chocolate, Gold, Gray, Lilac and Tortoise. This is one of the most popular and desirable show breeds. They grow rapidly. They are also raised for use in laboratories and for their meat.

Selecting Rabbits

Flemish Giant—The ideal buck will weigh fourteen pounds or more and the doe starts off a pound heavier. This breed has been known for several hundred years and is primarily raised for showing and for its meat. Recognized color varieties are Light Gray, Steel Gray, Sandy, Black, Blue, White and Fawn.

Florida White—Ideally both sexes weigh five pounds. This relatively new breed is cobby and is desirable as a fryer and a small laboratory animal. It is a white albino.

Havana—Ideally, bucks and does weigh six pounds. This animal has nothing to do with Cuba. It probably originated in Holland and was so called because it was the color of a Havana cigar. The colors recognized today are Blue and, of course, Chocolate.

Harlequin—Ideally both sexes should weigh between six and eight pounds. It was once known as the Japanese although it originated in France and has been exhibited there since 1887. This is a striped animal. The colors are applied on an orange or dilute background in the case of the Harlequin variety. The Magpie variety has a white background. Colors are Black, Blue, Chocolate and Lilac.

Himalayan—Bucks and does ideally weigh three and a half pounds. It looks superficially like a dwarf Californian. Old animals tend to fade and so they don't win prizes. This is an all-white animal with velvety black extremities. It would make a person think of the color pattern of a Siamese cat. The young start out silvery gray all over and gradually the body whitens and the points blacken. This is primarily a show and meat rabbit.

Lilac—Ideal bucks weigh six to seven pounds and does are a half pound heavier. This is a recessive derived from the Havana. The color is sometimes called pinky-dove. This is a good meat producer.

Lop—There are two recognized breeds of Lops in the U.S.A. The big floppy ears are their outstanding feature. The record spread is 28½ inches. Two feet of ear spread is common. Self colors and broken colors with white are all acceptable. Bucks ideally weigh ten pounds and does are a pound heavier. The breeds vary in technical details. One is called English and the other French; only experts can tell them apart. Both breeds have use in the laboratory and are also raised for showing.

Netherland Dwarf—There are 21 varieties of this tiny breed. Bucks and does ideally weigh only two

Selecting Rabbits

pounds, certainly not over two and a half. The coat is rollback and not flyback. This is the smallest rabbit breed and is primarily a show rabbit and a pet.

New Zealand—This famous great meat producer is also raised for laboratory use and for showing. Ideally a buck will weigh ten pounds and the doe eleven. The recognized color varieties are White, Red and Black. There have been New Zealands in the U.S. since 1912 and they have always been popular.

Palomino—Ideal bucks weigh nine pounds and does a pound more. There is a Golden-Lynx variety which is recognized by the A.R.B.A. This is a meat rabbit colored light golden with creamy undercolor. The Golden-Lynx has a silver sheen on an orange base color. There is light lilac ticking.

Polish—Another dwarf breed, this one is ideally two and a half pounds for both sexes with an upper limit of three and a half. There is probably nothing about it to tie to Poland. They have been exhibited

Townsend's Hares.

Selecting Rabbits

in England for over 100 years, and they are also bred for use in laboratories. Color varieties are Black, Chocolate, Blue and Ruby-Eyed White.

Rex—The ideal buck weighs eight pounds and the doe a pound more. The color varieties are Black, Blue, Californian, Castor, Chinchilla, Chocolate, Lilac, Lynx, Opal, Red, Sable, Seal and White. This breed is a mutant which showed up from time to time but was ignored until 1919 when a French breeder began to promote them. The distinctive feature is that there are no conspicuous guard hairs. Guard hair is present, but it is the same length as the undercoat. The fur resembles plush; and it is for its fur, and for show, that this breed is raised. This breed is also a good meat producer.

Sable—There are two A.R.B.A.-recognized varieties of this breed. The larger is called Sable and it ideally weighs eight pounds for the buck and nine for the doe. The smaller is called Siamese Sable and the ideal weight for bucks and does ranges from five to seven pounds. This is a mutant from the Chinchilla, which didn't come from South America, sometimes crossed with the Havana, which didn't come from Cuba. The Siamese, in case you wondered, didn't come from

Siam. The color is a rich sepia brown. The head, feet and ears have a purple sheen. The flanks and belly are paler.

Satin—Bucks ideally weigh nine and a half pounds and does are ten pounds. This breed is recognized in Black, Blue, Californian, Chinchilla, Chocolate, Copper, Red, Siamese and White. Satins are a hair mutation which were developed in Indiana among some Havanas. The fur of the Satin is unique because of its sheen. The hair is thin and transparent, short, fine and dense. The guard hairs should be only slightly longer than the undercoat. This is a recessive genetic trait, but it does not affect meat or show qualities or other features of the animal.

Silver—This is ideally a six pound rabbit in various shades of silvery gray, fawn and brown. This breed should exhibit uniform silvering and ticking over its entire body.

Silver Fox—This rabbit was formerly known as the American Heavyweight Silver. It has relatively long fur which resembles the silver fox. Ideally the buck weighs nine and a half pounds and the doe a pound more. The official colors are Blue and Black.

Silver Marten—Ideal bucks weigh seven and a half pounds and does a

Selecting Rabbits

pound more. The varieties are Black, Blue, Chocolate and Sable. This is a Chinchilla sport. The British call it the Silver Fox Rabbit. It is primarily raised for showing and for its fur.

Tan—The buck, ideally, weighs four and a half pounds and the doe weighs five. This breed resulted from a cross between a wild English rabbit and a Dutch. By 1894 it was recognized in Europe. The color varieties are Black, Blue, Chocolate and Lilac.

Keeping Track

If you have two or three pet rabbits you might want to write down their birth dates and weights on a card and hang it on the outside of the cage. If you have a dozen, you have no choice. You must keep a written record or soon your breeders will be overage and your young stock will have gone to the pet shop or the kitchen stove. It is amazing how fast we forget, especially if they all look much the same.

It is also true that most of us would have trouble picking a particular individual out of a herd of a dozen albinos. A little girl with her first pet rabbit would never

believe that her animal looks like all the others, but it really does.

At a show, rabbits are marked with a felt-tip inker for quick identification, but the "legal" permanent marking consists of a tattoo in the ear and/or a metal leg

Captions
Page 17, all Netherland Dwarfs. top right: Smoke Pearl Siamese; top left: Dutch Brown; bottom: Agouti. Page 18: top left: Chinchilla rabbit; top right Chinchilla Netherland Dwarf; bottom: Black Silver Marten. Page 19: top, Black Rex; bottom: Blue Silver Marten. Page 20: all Netherland Dwarfs: top left: Blue Silver Marten; top right Black-marked Himalayan; bottom: Himalayan-marked Netherland. Page 21: top, Blue or Gray Dutch; bottom: Chocolate Dutch. Page 22: top right: Blue Silver Marten Netherland Dwarf; top left: Brown Silver Marten Netherland Dwarf; bottom: Brown Satin Silver Marten. Page 23: Two Chinchillas. Page 24: top left: Florida White; top right: Ruby-eyed Netherland Dwarf; bottom: Ruby-eyed White Rex. All color photos by Michael Gilroy.

19

Handling Rabbits

ring on a hind leg. The inscription of the tattoo is assigned by the breeders association you join, and you *should* join.

The British Rabbit Council offers nine sizes of leg rings for its members' use. The numbers are recorded for each purchaser; and if a ringed rabbbit is sold alive, the name of the new owner should be filed with the council. The ring is usually slipped over the animal's hock when it is about ten weeks old and remains on for the life of the rabbit.

Some people have used ear tags on rabbits, but I don't recommend them. They tend to tear out or hang up on hardware in the cage.

Part of keeping track should be a record of the weight of the animal accurate to one or, at most, two ounces. For this you should have a beam balance. A good new one is expensive, but a good used one sometimes shows up in a sale of household goods. It is generally fashionable to weigh babies frequently at home—a large old rabbit weighs about as much as a small young baby, so this is the type of scale you should watch for. Unfortunately, most spring scales are not precise enough for our purposes in the rabbits' range of weights.

A tame rabbit seems to enjoy being handled by its owner. A nervous or frightened animal or a tame animal in the wrong hands can lacerate a person or kill itself as it struggles to escape. Since rabbits range in weight to as much as fifteen pounds, you must know what you are reaching for and act accordingly.

If you are going to own a rabbit, you must learn how to properly hold it so neither you nor the rabbit are injured by the experience. A two-pound rabbit must be held differently than a 15 pounder. Tame rabbits enjoy being held.

Handling Rabbits

Holding a small rabbit in your palm.

A two- or three-pound adult dwarf or a young meat animal can be grasped firmly but gently across the loins if your hand is large enough. Experienced handlers do it all the time and never hurt the animals; but until you have had some practice, use two hands.

With one hand, gently grasp the ears just above the skull in order to steady the animal. Do not lift it by the ears, *ever*. With the other hand, support the animal as you lift it with your palm under its chest, fingers on one side and the thumb on the other side.

A large rabbit should be lifted with its head up and its belly against your chest. One hand should support the back of its neck; the other hand should be under the rump to carry most of the weight. This cannot be accomplished safely unless you are wearing a very heavy canvas apron or smock. The claws on the hind legs of a large rabbit can easily tear apart another rabbit and can certainly rip a person's clothing or worse. Some rabbits also bite. Don't get the idea for even a moment that every rabbit is just a darling doll-like Easter Bunny. The hind legs of a rabbit are dangerous weapons and on occasion have struck out with such force that the poor animal broke its own back!

Larger rabbits may require the neck and the bottom held.

26

Housing

Many books about animals tell you to build the house and then buy the livestock. This is a universal truth, but with rabbits you must take one additional step. It would be a mistake to build before you have decided what you wish to accomplish. One reason is that rabbits range in weight from two and a half pounds to over fifteen pounds. There are about 30 breeds recognized by American rabbit fanciers and more than 70 varieties of those breeds are available. Rabbit fur ranges from that of an Angora, several inches long, to that of Satins, whose hair is not only extremely short but is so flattened as to make its insulating value practically nil. Some varieties are known to be hardy and others are considered somewhat delicate. Obviously, housing for a hardy ten-pound New Zealand Red will not be the same as for a Satin or a Netherland Dwarf that weighs only three pounds. You don't need to invent any rabbit housing until you have seen and used what others in your area have already settled on. I suggest that for the next ten or twenty years you experiment with established cage designs before doing something new. Now, to get more specific, here are some guidelines to get you going in the right direction.

Look at what others in your area are doing. Ignore designs from distant places. A Florida arrangement is useless in New Hampshire.

Keep wood away from urine and droppings.

Wood must be covered with metal wherever it could possibly be gnawed.

Closures must be strong, rust-resistant and situated so that urine and feces do not pile up or corrode them.

Provision must be made for growth. A pregnant doe in eight weeks might equal eight rabbits each weighing four pounds and she could be pregnant again, and again and again.

Cage accessories must be strong, rust-resistant and tip-proof. A rabbit could die of thirst over a weekend because it kicked over a water bowl on Friday. Of course it will have learned its lesson, but how about you?

Wood preservatives must be applied with understanding and caution. Many are poisonous.

Protection from rain and snow and biting insects and vermin and dogs and predators and undisciplined children and thieves must be considered in any cage

Housing

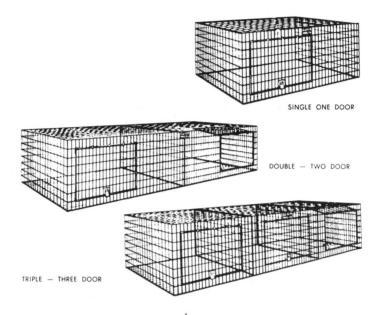

SINGLE ONE DOOR

DOUBLE — TWO DOOR

TRIPLE — THREE DOOR

design.

A metal-screen cage bottom is great for ventilation and getting rid of fecal pellets and urine; but if you have a thumper, the poor animal will soon suffer from sore hocks as a result of the hammering.

Another problem with an all-metal cage bottom is that if the cage is located outdoors in a cold climate, the metal will quickly carry away the body heat of the animal. A wooden board perhaps ten inches by eighteen or smaller for a dwarf variety might easily prevent a dangerous chilling.

There are two simple but extreme methods for housing your animals.

These Glick cages can be installed in tiers, one above the other. They are called single, double or triple doors. They are excellent and clean.

One is to release a few on an island where there are edible plants and few or no predators; an island like Australia would be good, for example. The other extreme is the controlled environment method where feeding and even cleaning of cages is timed and even automated. This latter approach is typified by some poultry plants where everything, even the egg, moves on conveyor belts.

28

Housing

Somewhere between these extremes you will settle on the arrangement that best suits your situation. What must you know before you get started? Now let's get even more specific; just a few simple numbers and a few basic decisions are required.

First: A caged adult doe should have one square foot per pound of her live weight. Dwarf varieties should have no less than four square feet of floor space no matter how small the animal.

Second: Young rabbits kept together when under eight weeks of age and under four and a half pounds of weight should be given no less than two square feet of floor space per animal. Thus, an average litter of six or seven could do well in a cage six to eight feet long and two to two and a half feet wide.

Third: A single buck used for breeding needs more room per

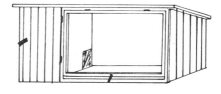

A single rabbit hutch, above. A movable cage can be used to put the rabbits where the grass grows, or it can be attached to a hutch, below.

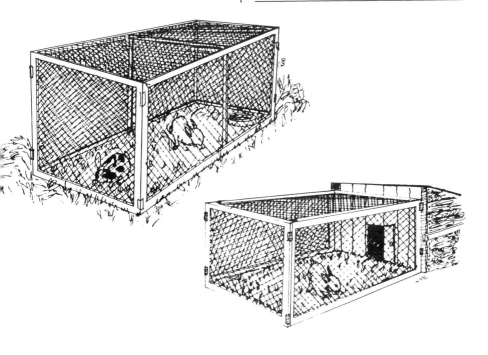

Housing

pound than a comparable doe so give him a cage based on two square feet per pound.

Fourth: The height of a cage depends on the size of the largest animal you plan to keep in it. Two feet is a good number for a six-pound rabbit. Dwarfs could manage nicely with eighteen inches and giants should get 30 inches. Cage height is important for ventilation and security as well as creature comfort.

Fifth: The rabbit got its scientific name from living in a cave or tunnel. Don't try to remake it. Give each adult animal a private place. This is especially important for a breeding doe. A nest box should be attached to any cage where a doe is kept for raising her family. The nest box should be small enough to suggest an underground place of refuge but large enough for the doe to nurse a litter of young. A small doe might do well in a nest box measuring twelve by twenty by twelve inches high. The entry should be large enough for her to pass through without scraping and high enough to keep the newborn young from creeping out. The top should be rabbit-secure but removable-by-you for examination and cleaning.

Rabbits are accustomed to close quarters, for they are, after all, cave or tunnel living animals. This hutch is meant for inside the home. It's ideal for one or two rabbits of medium size.

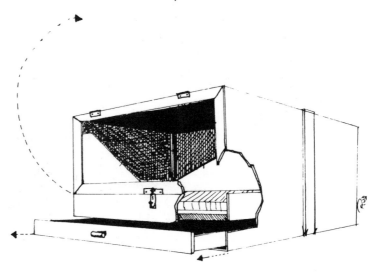

30

Housing

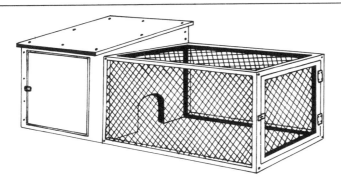

If your final rabbit housing arrangement smells musty or smells like ammonia after the animals have settled in it, you are doing something wrong. A well-designed and well-cared-for rabbit establishment for one animal or several hundred should be virtually odorless even on a hot summer day.

Now, this was mentioned before, but it bears repeating. Before you build or buy cages, do some research. Visit nearby rabbit keepers; see how they do it. Rabbits spend generations of lifetimes choosing and modifying their living quarters. If we are to do it for them, it behooves us to do it thoughtfully. Subtle things like drainage, prevailing winds, the angle of the sun, predators, pests and zoning regulations will all deserve consideration. Even if your goal is to maintain a herd of 100 breeders, you should start with no more than three or four animals and you

This hutch and cage can be made as large as required, but it is best that only one pair be kept in it.

should keep them through a couple of breeding cycles before taking the next big step. You may discover that you have a fox or a Tasmanian Devil or an irate neighbor lurking in the bushes. You might also discover that some member of your family is allergic to fur.

One important but occasionally overlooked detail about housing your rabbits is that if you start with a buck and three does and you keep their progeny, you will be the master of about 150 animals in the second generation—all before that first year has gone by. In addition, some does from those first litters will be reproducing themselves. By the end of the second year, if all goes well, you will be under the control of over 5000 rabbits. Plan ahead.

31

Housing

This is a home-made cage that is suitable for rabbits, hamsters, guinea pigs and the like. The 1" box screening is heavy and galvanized and is attached inside the wood so the chewers can't gnaw their way out of their own hutch!

If you elect to build your own rabbit housing, I suggest that you begin not only with a proven design but also by using materials which have been proven effective over the years. Screening is the item where most novices get into trouble. They go to a neighborhood hardware store and end up with chicken wire. It is called chicken wire because it is used to enclose chickens, and for that purpose it is great. Rabbits need a rectangular mesh with approximately five-eighths to three-quarters square- or half by one and a half-inch openings of sixteen-gauge steel wire or even heavier material. It should be galvanized after weaving or after welding. It should make you wonder why it needs to be so heavy. The answers are that most cage designs call for wire bottoms which must support not only the weight of a doe and her litter—this can amount to about 40 pounds—but also a bit of thumping from time to time. Additionally, rabbit urine will tend to attack metal. Galvanizing does help, but it will not add strength. Also there is the matter of security. A small dog, an opossum or a raccoon can tear chicken wire and wipe out your entire herd in a matter of minutes. What isn't killed might be frightened to death; the effect is the same.

The mesh size is intended to permit passage of feces. The heavy gauge will permit you to clean the cage quickly and efficiently. You simply remove the animals and play a propane or LPG flame over the entire structure from time to time. This gets rid of not only adhering waste materials but also hair, cobwebs, disease-causing organisms and insect pests. Even a wood-frame

Cage Furnishings

hutch with heavy mesh can be flame-treated a couple of times a year with no damage, but a flame-treated chicken-wire enclosure would not do as well. If you need screened cage-covers which are out of harm's way, you might use galvanized chicken wire for that special job; but frankly, for the long haul I don't think it is worth what you will save on the initial cost.

Start slowly, start small, start by subscribing to a rabbit breeders' magazine, join a rabbit breeders' association, visit a few good breeders, don't under any circumstances design a new rabbit hutch. If you must invent something, try for a wheel.

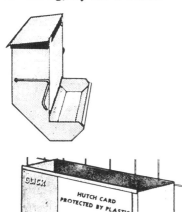

Essentials and Nonessentials

You can buy an exhibition-quality buck and two does, registered pure-bred stock from prize bloodlines of many breeds, for less than a hundred dollars. In fact, there are a lot of youngsters from good breeders available for fifteen dollars each. In fact, if you beat the butcher to the hutch you can buy meat breeds for what they sell for by the pound, which is certainly less than a dollar a pound. At that rate, a trio of five-pound rabbits would weigh fifteen pounds and would cost you less than fifteen dollars.

Now, to house these animals you can easily end up on a roller coaster if you are careless or have an uncontrolled credit card. First off, there are some things you must have. You will need a cage for each animal as soon as it is sexually mature. If you buy these cages you are looking at spending as much as fifteen dollars each—unfurnished. Even if you are a competent mechanic you can't build them for

This Glick feeder attaches to the screen on the cage. Wire cutters are necessary to attach it. A plastic identification card comes with it.

Cage Furnishings

much less. You will still need the clip pliers, another added cost, to close the "J" clip used to fasten one piece of cage mesh to another. Wire cages, incidentally, are pre-cut and shipped knocked down. You assemble them. The day when a person could build rabbit housing from packing crates is over.

Rabbits must drink and your local petshop will be able to supply you with ready-made water bottles which offer your rabbit water upon demand.

Wood crates went the way of the Railway Express Agency and the "lucky stick" in a Good Humor ice cream bar. The town dump of yesteryear furnished all but the nails to house rabbits. Today it is called a sanitary landfill and a person is subject to arrest if caught even looking at something to salvage or exchange. So you pay.

Now, in addition to the bare cage you *must* have some provision for water and food and a nest box for each doe. And you should become a member of the A.R.B.A. These then are absolute "musts." Additionally, and here is where I warn you to start slowly, there is equipment available for artificial insemination, heating pads for newborn young and, don't forget, tattoo equipment.

A tattoo is required if you show your rabbits, but perhaps you could find someone who just retired from rabbit keeping or someone who will let you borrow the equipment—a new complete first-class outfit could cost as much as 70 dollars.

Cage Furnishings

Then there are automatic water dispensers, medicine and food supplements and a carrying case for fifteen or more dollars. Hoppers for rabbit pellets can cost several dollars each, if you can't build them; and water bowls cost almost as much, if you cannot find something to make do. A plastic rabbit nest nineteen by fifteen by sixteen inches high will probably cost over ten dollars (this is something you could build or maybe arrange to buy from your local high school woodworking shop).

If you slaughter your animals, then you will need knives, hide stretchers, skinning hooks, and perhaps a taxidermy book.

A spring-scale with a scoop might be had for about twenty dollars and a good beam-balance could easily go for 75 dollars.

Then you could buy hutch cards and pedigree blank forms and a magazine subscription. And then there is membership in an A.R.B.A. specialty club. These are all worthwhile but not essential for getting started.

Too many people have had their enthusiam for this really wholesome avocation spoiled because they got carried away with the dry goods and they lost sight of the animals.

If you are going into rabbits for fun—and that is really the only way to get started—you can make do nicely with rabbits, cages, pellets, hay and those few additional basics which I listed at the beginning of this chapter. The heating pad for the nest is probably worthwhile if you want production in February and you live in Wisconsin or North Dakota or northern Maine. But really, must your does kindle in twenty below zero weather? And must you know exactly what they weigh? And must you bring your first litter with its ears tattooed to a sanctioned show?

So, in my view you should start your rabbit experience with just the basics. Period.

Oh, yes, you could also get some special rabbit nail clippers and a nice heavy-duty white smock or a laboratory coat. Perhaps this could be a reward to yourself if you are still keeping rabbits at this time next year.

Bedding

Rabbits in heated buildings don't absolutely require bedding. Some laboratory research would be impossible if rabbits had cage bedding (also called litter) in their cages, but in my opinion all other

Cage Furnishings

rabbit housing should employ litter. Certainly, litter should be placed in the nest boxes of pregnant does.

The reasons for cage bedding are worth a short review. To begin with, a caged rabbit really doesn't have much to do during its waking hours. If the bedding is edible, the animal will enjoy nibbling at it even if it is not especially nutritious. Also, litter is a great absorber of moisture and odors. Of course, it is then incumbent on you to clean the cages and replenish the material before it becomes damp or moldy.

In the nest of a doe about to give birth, bedding is a cushion which will protect the young; and if it is piled in the corners there is less risk of losing the pup who wanders off and gets chilled.

If the bedding is palatable, even if it is not nutritious, your animals will work on it instead of on the wooden parts of the hutch. A rabbit needs roughage in its diet—and somehow, somewhere, the animal will find it.

In cold climates, the bedding in a nest box is an excellent insulator; an armful of hay or straw could easily make the difference between disease and robust health.

Which litter or bedding is best? I prefer timothy hay because, although it is not the best absorber,

it is avidly gnawed and it is somewhat nutritious. Second choice would be barley straw or a bright yellow spring wheat straw. It is softer than timothy but also less nutritious and less expensive. Rye and oat straw are quite rough and not desirable. Other bedding materials which have been used in various places at various times are dried hardwood sawdust, wood shavings, wood chips, peat moss, strips of newspaper and even tree leaves. Avoid laurel leaves since some species of laurel are poisonous.

Torn or shredded newspaper is frequently used as bedding. You may not see anything but newspaper in pet shops. It is absorbent but offers few other advantages. The black ink on newspaper will rub off and make a white rabbit look gray.

Wet or dirty bedding is self defeating; it attracts flies and other insects, it causes or contributes to disease and it generates unpleasant odors. Fortunately it is not very difficult to keep rabbits clean and odorless. They want to be clean. They will defecate in the same place, regularly. There are many accounts of rabbits kept in homes, uncaged. These animals would use a tray as reliably and fastidiously as any house cat. The problem (as with cats) is that a buck (like a tomcat) will spray to mark the territories under his jurisdiction. A pet rabbit kept in your house had better be a doe.

Diet

The best ration for a captive rabbit is commercial rabbit pellets and clean fresh water. This is a safe, complete, convenient and reasonably inexpensive diet.

Rabbits don't like sudden dramatic changes in their food. Find out what they were eating when you acquired your animals and if you must make changes, do it gradually. It is not enough to know they were eating "pellets." Find out what brand of pellets. If possible bring a sample home with your new animals.

Store your rabbit food as if you plan to eat it yourself. Keep it out of the sun. Keep it dry. Keep it cool. Don't let mice get into it. Do feed it from a dispenser that cannot be wet, soiled or spilled.

This is a short simple chapter about a most critical aspect of animal husbandry; don't gloss over it. Many rabbit ailments are direct or indirect results of poor food or poor feeding practices. The rabbit is a fastidious eater and it cannot tolerate the abuses that we

administer to ourselves and our dogs and cats.

A rabbit pellet should be a scientifically compounded blend of hay, clover or alfalfa, grains and minerals. It should be relatively high in protein—read the label. The crude protein *must* exceed twelve percent and *should* range between sixteen and twenty percent if you are feeding to pregnant does and to does that are nursing youngsters. If you use a high-protein diet for all your stock, none will suffer. A five-pound rabbit will eat five ounces of pellets and maybe eat or wear out a half pound of hay per day.

The other diet element is water. You should furnish one pint per day per each four-pound animal. The water should be *at least* as clean, cool and fresh as what you drink. There will always be a problem delivering water. Rabbits upset the dishes, dribble the water on their fur, defecate in it, dump food in it, shed hair in it and chew up the plumbing. Also, the water freezes, transmits disease germs (even upstream in low-pressure delivery systems) and attracts insects which drown in it. There is no universally accepted, perfect, foolproof, inexpensive watering system for rabbits. For 2000 years, people have struggled with this problem and

Cage Furnishings

even after inventing television, computers and rockets to the moon the perfect rabbit watering system remains unsolved. As you get started, remember that a rabbit cannot get the water it needs from ice. The water must be in liquid form.

You will probably supplement the pellet-and-water ration with hay, straw and fresh vegetable treats. If you are introducing any new food, start slowly with just a taste on the first day. Your animals will enjoy freshly mowed lawn grass if it hasn't been treated with herbicides or insecticides. They will probably not eat citrus fruits, bananas or tomatoes. Youngsters will do well with fresh milk soaked up by dry bread. Just remember that the breed must not be moldy.

Rabbits are known to enjoy cabbage, apples, rutabagas, potatoes, Jerusalem artichokes, carrots and, if you can afford it, lettuce. Lettuce, incidentally, is probably the least nutritious of all the foods that your rabbit might eat.

The Dwarf rabbits shown below are called Agouti, or their natural wild coloration. Their feed needs are obviously much less than larger rabbits require. Give your rabbits all that they need—but not more.

Cage Furnishings

A salt block is necessary if you feed more hay than pellets—the pellets have salt in them.

One aspect of rabbit diet is unusual and might surprise you. This process is called coprophagy, and it explains why a rabbit is so efficient as a producer of meat per unit weight of food.

The digestive process of a rabbit works like this: The animal forages outside of its burrow as quickly as it can. It swallows much of its grassy and leafy food without thoroughly chewing it. This much is akin to the method of cattle and deer. Now, cattle and deer store their food in one of several stomachs and then regurgitate it at leisure to re-chew and then pass on to a lower part of their alimentary system to complete the digestive process. That is, they chew their cud.

Now, a rabbit passes some of what it eats all the way through its digestive system and it produces (from one complete pass-through) a soft "fecal pellet" which it quickly re-ingests and more completely digests on the second pass through its digestive tract. After the second pass a hard fecal pellet is excreted and this is what we see. Many rabbit keepers have never witnessed the eating of soft pellets—it happens so rapidly and usually is done while a rabbit is in its shelter, nest box or covered hutch.

If an animal stops eating for no apparent reason—generally there *is* a reason and you should patiently search it out—try tempting it back into the eating habit with a treat. Try a bit of carrot or leafy vegetable or apple or a piece of toast soaked in milk. One cause for going off feed is a hairball lodged in the digestive tract. Grooming during molting will help. Clean cages will help. Bulk in the diet might assist in carrying the hair through the tract. Gentle palpation might break a stubborn hairball loose. You might have to sacrifice the animal if all else fails.

39

Breeding

Sexing

The most experienced rabbit keepers will admit that once in a while they have been fooled; usually when this happens it is a young male that looks like a female. What's the difference, you ask? Plenty. Since it takes only one male to service as many as ten does, the surplus young males are generally sold to pet shops or to meat processors. Does are sometimes kept long enough to evaluate their potential as breeders.

If you hold weaned youngsters together until they reach say four and a half pounds, you may discover that a young buck has fertilized one of the does he has been caged with.

She will almost certainly resorb or abort or produce stillborns or perhaps cannibalize her litter—when a doe is too young these things are more liable to happen. So you should examine young rabbits and try to segregate the sexes by the time they are weaned. What do you look for?

Males have a sheathed penis which will protrude from a *rounded* opening. A female has a slit-shaped opening in approximately the same place.

Don't rely on finding a scrotum or testicles on a young male rabbit; they may still be in his abdomen, undescended.

All other things being equal, a male has a good chance of having a larger head than a female littermate, but don't count on it.

To determine the sex of a young rabbit, roll the animal on its back and hold one hind leg with each hand. You will be able, with practice, to use your thumbs to press gently on the lower abdomen in order to expose these organs.

Basics of Breeding

This section could be made to fill the entire book. Let me tell you the essentials here and now. The details and explanations can come after you have accumulated some experience.

Start with young but mature, healthy but not overly fat breeders. A small buck should be good from the time he is six or seven months old until he is four or five years old. Larger breeds mature more slowly. A doe of a dwarf variety might be sexually mature at four months but might make a better mother if she were not bred until, say, five or six months. You will be a better judge of your animals only after you have watched them for a few years. Some strains of the same breed exhibit

Breeding

wide variation in time to mature. Regardless, few does are worth breeding after their fourth year, and many fizzle out after only three.

The animals should ideally have been in their cages for a few months before being bred. They should have settled down and become accustomed to having you handle them. The doe should have had her toes checked. If the nails seem long, clip them to perhaps an eighth inch longer than the quick before mating her. An Angora's hair might need to be clipped, but you really shouldn't be breeding Angoras for starters.

Move the doe into the buck's cage. If she is receptive, copulation will take place in a few minutes and then the buck may fall over on his side. Don't worry—he *will* recover. If things don't work out right wait a few days, keep the cages close enough so they can smell each other, subdue the light, and keep things quiet.

The buck might be a bit feisty and you should maneuver to avoid being sprayed. The smell of ammonia is reason enough for you to be cautious.

After what seems to be a consummated mating, return the doe to her cage and let the buck rest for a week. Record the date of the mating on a cage card and wait a

When you squeeze a rabbit, as shown above, your fingers can put enough pressure on the abdomen to cause the penis to protrude. The two small drawings show a normal pore and one with the penis protruding.

Breeding

month. She will, if all goes well, deliver a litter of blind and hairless pups anytime from 28 to 35 days later (but usually on the thirty-first day).

Be sure she has plenty of fresh food and water, although she may go off her food a day or so before giving birth. She will probably pluck some fur from around her teats to mix with the bedding.

The doe may deliver a litter of any number between one and twelve pups. The average for a small doe is probably five or six and a larger animal might come through with six or seven on average. A doe will be unable to raise more pups than she has functional teats with which to nurse them. If an oversize litter is in the cage, you might be wise to remove any deformed or runty individuals and give the others a better start.

When pressure is put on the stomach of a doe, her anal pore changes shape, but there is no protruding penis.

Breeding

If you have two does that both kindled within three or four days of each other, you might try to foster an overproduction from one onto the other. Sometimes it works. Don't look for opportunities to manipulate your animals, but if you are in trouble it might be worth the risk. If you wish to switch youngsters, rub some damp bedding from the new cage over the orphan before you introduce it. If you have lost a doe and are stuck with no foster mother you should cull the litter. Hand-feeding is not something for a novice.

If you were hoping for Production with a capital *P* and you are disappointed, there are several details to check.

Are your animals well fleshed but not fat? A fat buck is not likely to be an avid and aggressive breeder. A fat doe might be impossible for any buck to inseminate.

Have the does been stressed? A pesty kid next door or a little dog or a raccoon at night might be the culprit and you not wiser that your does have conceived and then resorbed their fetuses.

Is your buck potent? If you just got through a hot summer your animals likely did too. A temperature of 90°F. or over for even a few days can render a buck impotent for several months. An older buck is more vulnerable to potency loss (impotency) than a younger buck if heat is the trouble.

Do your animals have sore hocks? A doe with sore hocks will be in pain when a buck mounts her and she may not conceive. Check both prospective breeders for this and all other signs of health. Your animals should be perfect if you want success.

Have your Angoras been clipped? Sometimes a long-haired animal cannot breed for this reason; certainly it is easy to correct.

Is your old buck worn out? A good buck can easily service three and ideally service ten does, but he cannot do this forever. If his present successes don't measure up to his past records perhaps he should be retired, full of honors. This points up the importance of keeping accurate records.

If a pregnant doe is stressed by fear or rough handling or lack of food or a chill she might abort or if she has not been pregnant for long, she will absorb the embryos and fool you. The process is technically called resorption. You might call it "false pregnancy" as though she tricked you. You could call it anything you please to call it but that won't alter the facts. In fact,

Breeding

you were probably in a position to keep it from happening. False pregnancy or resorption is less liable to happen in herds belonging to experienced rabbit keepers. This suggests human influence, doesn't it?

Rabbits and cats can be friends. Cats can be housebroken and so can some pet rabbits.

The rabbit is an efficient animal. By absorbing the unborn young, the doe conserves energy which would be lost if she aborted or if the young went to full term and were stillborn.

Another example of this efficiency is the way a doe plucks fur from around her teats before she kindles. She not only makes it easy for the pups to get to the milk but she also uses the wool to line her nest and insulate the newborn youngsters.

Breeders' Timetable

Mate a receptive doe anytime if she is more than six months old.

She will ovulate after she mates. Most other animals ovulate first and then mate.

About 27 days later she will pluck fur from her belly and make a nest.

About 28 but usually 31 and in never more than 35 days after mating she will give birth (kindle) to a litter of as many as twelve but usually fewer blind, naked pups.

Babies' eyes will open in twelve to fourteen days.

You may mate the doe again when the litter is about fifteen to twenty days old.

Young may be weaned when they are four or five weeks old as they gradually go onto solid food, but they do better if they stay with the doe for six to eight weeks.

Young are ready to be selected for breeding or for slaughter when they are eight to ten weeks old.

A doe can produce five or even six litters of six to eight young per litter per year.

She can produce for perhaps three years at this rate.

She might live five years on average.

No rabbit lives much longer than twelve or thirteen years.

Breeding

A meat breed doe might produce 75 to 100 pounds (live weight) of offspring per year.

One buck can serve as many as ten does and even so he might live out his appointed twelve or thirteen years.

Rabbit Colonies

You may read or hear about a shortcut breeding technique called "colonizing" or "colony breeding." Don't do it.

A colony is enclosed in a fence and it cannot spread out as it naturally might. The animals will be in contact with each other's urine and feces. Diseases will spread like wildfire.

A junior doe will be bred at three months instead of five and she will lose her babies.

A young buck will be torn up by an old buck or an unappreciative doe.

Don't colonize your rabbits; look what happened to Australia when their rabbits got outside the fence.

Genetics

Should a rabbit breeder understand genetics? Yes. Should you master this science before you start a breeding program? No. I suggest that for now you learn a few simple terms and words and the basic principles; then as you see the results of your breeding efforts you can relate what you see to what you know. You have time later to decide how much more you need to know. There are successful rabbit breeders with no formal understanding of genetics and then there are Doctors of Philosophy who have written scholarly dissertations on rabbit genetics but have never carried a dozen does through a good year of production.

Here are a few terms for you to get to know because other breeders use them. Many conditions have several names—simply use the one you feel more comfortable with.

A typical *recessive* trait is albino (white fur and pink eyes). This is the classic example of a simple recessive. If albinos from the same strain are mated, all their offspring will be albino. This is the surest and most restrictive application of genetic theory.

Pure, also called *homozygous,* means that there is no hidden genetic material. The animal is genetically what it looks like on the outside. It *might* be a pure agouti. If it is albino it *must* be a pure albino.

Breeding

Split, also called *heterozygous* means that there *is* a hidden genetic trait and this trait *might* become apparent in a succeeding generation. An albino cannot be split. As I stated previously it must be pure.

Dominant is a trait which is stronger than some other related trait. For example, the agouti pattern (wild rabbit color) is dominant over albino (white with pink eyes). Now the plot thickens. Every cell in every living creature contains microscopic chromosomes which carry pairs of sub-microscopic genes, and these genes are usually classed as recessive or dominant. A dominant gene masks a recessive gene and so an agouti rabbit might look that way because it has a pair of dominant agouti genes or because only one dominant agouti gene is present and it masks its paired recessive albino gene.

Now, let's take just one more step. If a doe with an agouti pattern was split (heterozygous) for albino then she would have a 50-50 chance of passing on an albino gene to her offspring. If that doe was mated to a buck that was also agouti but split for albino, then that buck would also have the same 50-50 chance of passing on an albino gene to his offspring. So what do we have? Let's look at a little chart which will display graphically what was just stated in words. The split agouti/albino doe is represented by two genes: an agouti gene (*A*) and an albino gene (*a*). Call her *Aa*. The buck is also the same, genetically speaking, for this color trait. Call him *Aa* also. Now mate these animals:

		Doe	
		A	a
Buck	A	AA	Aa
	a	Aa	aa

If you had a great number of offspring they would appear in the same ratio as their graphical display:

AA   Aa   aa   Aa

There are four offspring representing three possible arrangements of genetic material.

Of the four, one or 25% is *AA* or pure, homozygous agouti. There is no albino genetic material here.

Of the four, two or 50% are *Aa* which is a split or heterozygous individual which looks like agouti because that is what is dominant.

Rabbit Products

Of the four, one or 25% is *aa* which is pure recessive albino. There is no agouti genetic material in such an animal.

Those then are the odds, but it doesn't mean that in every litter of eight there will be two albinos, two pure agoutis and four agoutis split for albino. It doesn't mean that any more than that every family with two children will have one boy and one girl. All you know is the odds, but you don't have any assurances on the short haul.

This then was your simple introduction. Rabbit genetics have been researched and written about for many years. There is plenty of literature on this subject in your pet shop or in your local library—how far to go is up to you.

Rabbits cannot climb to reach water. Their water must be on the ground at a very low level.

Meat

All domestic rabbits are edible. They all taste about the same. Certain breeds produce meat more efficiently than others, some are "meatier," some have smaller bones, some grow faster, some are the "right" size at the "best" age. But all are edible.

A meat rabbit that weighs about four or four and a half pounds when alive is most desired nowadays in the market place. This fits in nicely with the New Zealand and Californian breeds which will reach this weight in about eight weeks and will arrive at a rate of perhaps 50 per year from one good doe.

Now, a dwarf rabbit might weigh only three pounds when fully grown and a giant breed might have fewer offspring and a slower rate of growth; so if you think you would like to eat some of your surplus stock or sell it for meat you might bear this in mind when you choose a breed to keep.

When you are about to dispose of your animals you can sell them alive or you can slaughter, dress, chill and/or freeze the carcasses as simply as one would a chicken.

If you elect to do it yourself, I suggest the killing method recommended by the U.S.

Rabbit Products

Department of Agriculture in Handbook No. 309 entitled *Commercial Rabbit Raising.* This booklet is available from the Superintendent of Documents, U.S. Government Printing Office, Washington, DC 20402. This is the method as they present it:

The preferred method of slaughtering a rabbit is by dislocating the neck. Hold the animal by its hind legs with the left hand. Place the thumb of the right hand on the neck just back of the ears, with the four fingers extended under the chin. Push down on the neck with the right hand, stretching the animal. Press down with the thumb. Then raise the animal's head by a quick movement and dislocate the neck. The animal becomes unconscious and ceases struggling. This method is instantaneous and painless when done correctly.

If you wish to proceed with the skinning and eviscerating, you should get that government booklet or better, if possible, visit a hunter or another rabbit breeder who can demonstrate the procedure. Once you have seen it done you will find that it is not difficult.

Manure

Any rabbit fancier will tell you that the greatest thing for strawberries is rabbit manure. When I was a boy 50 or more years ago, the standard response coming from an insane asylum was "I use cream and sugar on my strawberries and everybody says *I'm* crazy." As a matter of fact rabbit manure is excellent for gardens and may be applied lightly at any time, with or without lime. Another proven application for rabbit manure is to raise earthworms in it for fish bait. When the worms are grown, the manure remains desirable for gardens and potted plants.

Tame rabbits quickly learn the trick of coming to you for food and begging for it on their hind legs.

Professional Rabbit Keeping

Planning Ahead

If you have a few rabbits and you are doing reasonably well with them, you will probably begin to think about taking a big step and going full time—professional. Well, this is commendable but unless you have several things going for you, your time would be better spent addressing letters and licking envelopes; and your money would bring you a better return if you invested it in a blue-chip stock.

So what are those several things? You will need land which is zoned for what you propose. There is no good reason for getting started with a zoning violation or some irate neighbors breathing down your neck. Ideally your rabbits will be virtually odorless, but somehow someone will detect an irritating whiff on a hot summer day and it is then that you will need to be on the right side of the zoning law.

The British have a law on their books which specifically protects a rabbit keeper so long as he doesn't create a nuisance, but in the U.S. there is no such special legislation.

You will need more knowledge and help than you can get from this book. Join the American Rabbit Breeders Association or the British Rabbit Council or one of their national specialist clubs. Read the guides, study the standards, attend meetings, go to shows, meet other breeders. People enjoy sharing their "secrets," and you will learn a lot and, hopefully, you will pass these "secrets" on in time.

What else? Start slowly. Assume that if you are able-bodied and willing to put in a full day, every day, you can maintain 600 does; but don't start by buying 600 hutches and 600 animals. No matter how well you handled the buck and his three ladyfriends, it is still a far cry from a full-time commitment.

Don't invent a new cage design. Adopt a good proven cage which works in your climate.

Plan to utilize the entire rabbit. To start simply, you might sell live animals at eight weeks of age. There are buyers who will pick up your production and haul it to a packing house on a routine basis.

Read the advertisements that appear in the British magazine *Fur and Feather* and the American publications *Domestic Rabbits* (the official bi-monthly organ of the A.R.B.A.) and *Rabbits* (the oldest monthly rabbit publication in the U.S.). They list buyers as well as sellers.

The U.S. Department of Agriculture offers all sorts of

Professional Rabbit Keeping

literature and guidance to rabbit breeders through county agricultural extension service agents. They may be found as simply as by looking in the telephone book or by calling the nearest agricultural college.

Economics

The University of California has been active in rabbit research for many years—in part because of the tremendous number of rabbits raised on more than 13,000 farms in that state. Los Angeles County alone produced more than four million dollars worth of rabbits in one recent year! A report from this university is summarized thus:

Average number of young per doe per year 26
Average production per doe per live weight per year . . . . 112 lbs.
Average labor in hours expended per doe and her offspring per year 6.4

The report also breaks down the costs about this way:

Food 54%
Labor 30%
Interest and Depreciation 16%

The income from meat was 90% of the total income. Pelts, manure and brood-stock sales made up the balance.

A partly automated and largely mechanized rabbitry can carry as many as 800 does with the labor of one person, but this is an exceptional case. The realistic figure is more like 350 and the minimum to support one full-time person would be 250 does plus the requisite ten percent of bucks and perhaps ten to twenty percent of junior does for recruitment. Of course the older animals that are retired are sold as roasters rather than fryers, but they will bring only half the per-pound price.

One more interesting number is furnished by the Albers Milling Company. It is 3.4—this is the pounds of dry food required to produce one pound of live rabbit. The study which gave us this figure started when the doe was bred. She was then weighed. All the pellets she and her offspring ate until they were eight weeks old were weighed. She and her pups were then weighed again. The gain in animal weight in pounds was divided into the total weight of pellets eaten. It took 3.4 pounds of pellets to generate one pound of live rabbit. The U.S. Department of Agriculture came up with a similar figure. It took them twelve and a half pounds to produce a four-pound fryer, including what the doe

Professional Rabbit Keeping

ate.

If you have champion stock to sell and if you have established a reputation for fair dealing, you *might* get 150 dollars or even more for a single outstanding specimen. This would be top-of-the-line, not just a good animal but a great animal.

If you have pure-blooded show stock to dispose of, you might get twenty dollars to 40 dollars per animal from someone who wants to get started with that breed.

If you have fryers for a processor and you deliver them, you will probably be paid 50 or 60 cents per pound, live weight. A white rabbit usually brings five cents more per pound because its pelt is worth an additional twenty cents at the tannery. The quality of rabbit meat depends on breed and age and weight—it has nothing to do with color of pelt. As contrasted to a chicken, all rabbit meat is white meat, even the legs and back, regardless of breed or color or age.

An old large meat-rabbit usually five pounds or more enters commerce as a "roaster" and it will bring eighteen to 30 cents per pound on the hoof delivered to the processor.

So you want to know how you can make money if you have a buck and three does and you produced 200 pounds, live weight, last year. Well, you might have to spend more than the 100 dollars for driving those animals to the slaughterhouse. Oh, yes, you can also sell the manure and the earthworms out of the manure. No, a meat operation with three does is a losing proposition. It is a matter of the economics of scale. No backyard chicken breeder can match Perdue or Holly Farms for price and no small rabbit keeper can enter the market place with meat at 50 cents per pound live weight.

Yes, there are some fairly large meat-rabbit establishments and some of them do survive. Pelts, manure and meat all add up to a reasonable profit for a dedicated and hard-working operator especially if the plant is ideally located for cheap food and easy shipping. Frankly, if I had money to invest in rabbits or in blue-chip stock, I would choose the stock. If I had less money but more time and labor to invest, then, and only then, would I go for the rabbits.

What thousands of people do accomplish with a lot of pleasure is to raise rabbits for fun and eat the surplus or sell them locally. If you start with meat producers of show quality you might make your

Professional Rabbit Keeping

expenses to the A.R.B.A. sanctioned shows. If you keep Angoras or some other specialty breed, the culls are still edible, while the exhibition stock might be salable for real money but don't hang your hat on it.

I already reviewed the costs for wire cages. If you elect to build your own you had better have a source of inexpensive lumber and some real skill or what you build will cost more than what you would buy ready-made.

Food and bedding are the next and only significant items. Your eight-pound rabbit will eat about a half pound of pellets per day or perhaps a little less if you offer something like a good bright timothy hay containing clover or alfalfa as a bedding and food supplement. So then the pellets for the year for one animal will weigh 365/2 = or about 180 to 200 pounds. In 1981 the local feed store price ranged around eighteen cents per pound in 25- or 50-pound bags.

At eighteen cents and 200 pounds per year we have a food bill of 36 dollars plus the cost of the hay. Say a bale costs four dollars, and for a rough approximation hay costs forty dollars per year per adult rabbit. This explains why a four-pound

rabbit at 50 cents per pound must be sold when it is only eight weeks old if it is to earn anyone any money. Of course a volume producer of rabbit meat will be buying pellets for a lot less than eighteen cents per pound and hay for less than four dollars per bale.

The rabbit-raising industry has been the victim of more quick-buck, confidence game and bunco artists than any other agricultural endeavor I can think of. Even today you will find all sorts of offers where you put up some money, you buy some animals and you work to raise these animals which someone promised to buy; and then when you have some animals, you wonder where that buyer went. He went to Rio or Buenos Aires or some little island in the Mediterranean to enjoy the money that he milked out of you.

Rabbit Health

Most books about rabbits have a chapter dealing with diseases. Perhaps a better approach, especially for beginners, would be to take a position of aggressive hygiene and nutrition, thus aiming to prevent disease rather than to cure it. In point of fact, many successful rabbit breeders will eliminate promptly and ruthlessly any chronically ill or defective stock from their herds in order to protect the others from infection or inheritable genetic disorders. One reason for taking this approach is that many rabbit diseases come on quickly, are hard to diagnose and are frequently fatal no matter what you do.

Signs of Good Health

When you begin your herd, start with absolutely healthy animals. Here are a few hints. Some appear elsewhere in this book, but this checklist may prove useful.

Erect or certainly unbroken ears (except in Lops, of course).
Bright clear eyes.
Clean anal openings.
Dry noses.
Clean dry dewlaps.

Straight legs and untwisted tails.
Alert but relaxed animals.
Clean skins, free of lumps, bumps, boils, blisters, cuts and bare spots.
Look at the cages they are coming from. Are they clean and dry?
Look at the food they were getting. Was the cat sleeping in the drum of pellets?
Was the dog in the hay?
Are there any signs of rat or mouse droppings?
Examine the rabbit's hocks. Are they fully furred and free of blood or soreness?
Look deep into their ears. Are they clean and free of scabs or matter?
Examine their teeth. Are they even and are the incisors gently curved to slide evenly against each other?
Hold each rabbit in your own hands. Does it feel solid and strong and meaty or does it seem weak and limp and bony or overly fat?
Cradle the animal in your arms. Does it bite or fight wildly? You don't want that.
If its toenails are too long but are not twisted, you can trim them when you get the animal home; but if you find the other defects I've mentioned, I suggest you leave these rabbits where you found them.

Rabbit Health

Don't let this checklist frighten you. For example, I attended a rabbit show recently where about 2000 animals were exhibited. I think I saw about all of them and none of these problems were evident. Many of those rabbits were available for less than 25 dollars. There are plenty of good rabbits available at fair prices.

Keeping Rabbits Healthy

Now, how do you *keep* your animals healthy? Again a checklist:

Quarantine any new animal for a month before you introduce it to your herd. A separate cage 100 feet away or in another room is a reasonable quarantine.

Sterilize the water supply or watering bowls or bottles frequently.

Store pellets and hay so that rodents, cats, dogs and birds cannot touch it. Tapeworms are transmitted from other animals to rabbits through fleas which get into the cages with food.

Protect the food supply against dampness. Digestive disorders follow eating damp or moldy pellets or hay. Furthermore, an animal run-down from this problem is more liable to contract still another infectious disease.

Don't crowd your animals. The formulas I gave you elsewhere are reasonable and have been proven by experienced breeders. Crowding is false economy.

Watch your animals carefully and should something go wrong, act promptly.

If you have a really valuable animal that sickens, be prepared to take it to a veterinarian and be prepared to pay his fee. He is a professional.

Watch your animals carefully in hot weather. Actually they can handle the cold better than the heat. If it is warmer than 90 °F. for any extended period, your bucks may become temporarily sterile. Ventilate with a fan. Spray water on hutch roofs. Place hutches under shady trees. Plan ahead if you live in a hot place. See how other rabbit keepers handle this problem.

Common Ailments

If your animals are not well, read over the descriptions of common ailments given here. I hope reading over the descriptions is not

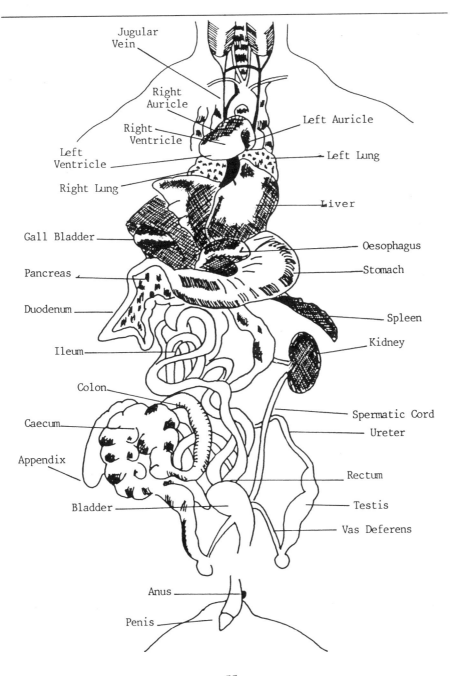

Rabbit Health

necessary. Many people have kept many rabbits for many years without losing any to disease. They may have been slightly lucky, but they were surely very thorough when it came to hygiene and good nutrition.

Sore hocks are caused by rubbing or thumping on a wire cage-bottom or on wet bedding. Some short-haired breeds are more vulnerable. Some high-strung thumpers are also vulnerable. Correct the cage-bottom problem and treat the animal with a soap and water washing followed by a mild antiseptic cream.

Pneumonia is a disease of run-down animals who have been dampened and/or chilled. The lungs become congested and the afflicted creature usually dies. There is no worthwhile treatment once the condition has advanced to the point where you can diagnose it. Don't try to cure pneumonia, but do take preventive steps through good husbandry. Pneumonia usually follows some other disability or disease.

Tuberculosis is another lung disease which you cannot cure but which can be prevented. Generally, healthy rabbits don't contract tuberculosis. A laboratory examination would be required to confirm its existence. This is the same disease as is found in people and cattle and it is transmissible through milk and contact. This disease is rare in rabbits, but people can be fooled by another bacterial disease which presents similar symptoms. Call it Pseudo-Tuberculosis. Again, breathing is labored and about a month after the infection, the rabbit suddenly dies. Small yellow-white nodules are found throughout the body organs. This disease is transmitted through infected food from mice and rats.

Captions
Page 57: top, Tan satin; bottom: Tan Rex. Page 58: top left, and bottom: Chocolate English Angora; top right, Black Silver Marten, also called a Silver Fox. Page 59: top, a pair of Black-marked Himalayans mating; bottom, a litter containing baby Black-checked Giants and Agouti Netherland Dwarfs. Page 60: a light Tan and a Red Rex. Page 61: top, two Black Dutch; bottom, an Agouti Netherland Dwarf. Page 62: top left: a Gray English Spot; top right: a Black Dutch; bottom, a Black English Spot. Page 63: top, young Gray English Spots; bottom, a Tan English Spot. Page 64: top left, a Black Netherland Dwarf; top right, Black Netherland Dwarf; bottom, Black Tan rabbit. All color photos by Michael Gilroy.

58

Rabbit Health

. *Myxomatosis* is a viral disease that is closely tied to the recent history of *O. cuniculus* worldwide and that does not affect other animals in exactly the same way. It is transmitted by mosquitoes, fleas and perhaps mites. Understand clearly, it does not affect humans. This disease seems to have originated in some South American lagomorph and it was taken to Australia in order to control the tremendous population of rabbits which were overrunning that country after having been deliberately introduced there about 100 years ago. Well, it did knock down the Australian rabbits, but somehow it was also accidentally carried some years later to Europe where it was previously unknown. Remember that *O. cuniculus* is a wild rabbit in much of Europe including the British Isles. It is not the only lagomorph, but it is a common and important one. Remember also that this selfsame species is the one and only domestic rabbit worldwide.

Well, myxomatosis went through Europe and Britain like a dose of

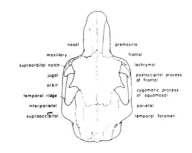

The Skull of a Rabbit. *Top view above; side view below.*

65

Rabbit Health

salts. It wiped out an important food supply for many people. It affected the fur and felt business. It worked its way to the U.S.A. where it became especially troublesome in California.

It causes a sleepy red-eyed rabbit. There may also be swollen eyelids. There may be lesions and swellings. Most of the affected rabbits die. There is no cure. They live long enough for the flea and mosquito and the mite to jump off and transmit the disease to still other animals. Those that survive become immune.

There is a preventative vaccination which may be effective but the best prevention is by knocking out the vectors—that is the creatures that transmit this viral pox. This means the use of sprays and screens and quarantines.

An interesting sidelight is that the Australian survivors are developing into a race of rabbits with an inherited resistance to this disease.

The Skeleton of a Rabbit.

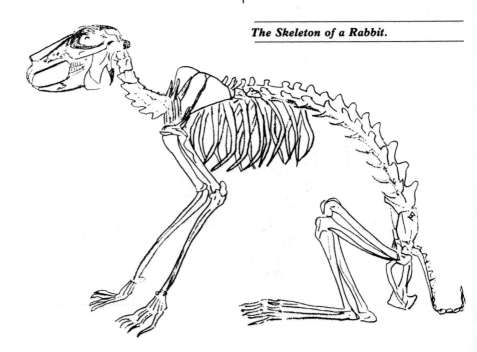

66

Rabbit Health

Tularemia is still another disease which you will probably never encounter but of which you should be aware. With your knowledge you can easily be protected. It has been found in North American wild animals including lagomorphs, rodents and birds like pheasants and quail. It is often fatal to the animal and it can be fatal to humans as well. It is rare in caged rabbits but relatively common in wild lagomorphs. It is transmitted by ticks and biting insects including flies. Mosquitoes also transmit it. A wild cottontail rabbit will be dead in a week. Before it dies it will be sleepy, depressed, huddled, rough-coated and stiff.

A hunter is well advised to avoid a "sleepy" cottontail or jackrabbit. When any wild rabbit is dressed, the person doing it should wear blood-proof gloves. This disease is one good reason why you should not let your animals get close to any other animal, especially wild rodents and wild rabbits or hares. Most rabbit keepers will go through a lifetime without encountering this disease; but remember, one dose will suffice.

Coccidiosis is the last disease you need to know about as you get started. Some large-scale commercial breeders control it with antibiotic medicine. Coccidiosis has been known for a long time; and interestingly, this is the first identified protozoan disease! The disease-causing organism was first seen and described by the pioneer of the microscope, Leeuwenhoek, in 1678!

This is a disease of malnourished rabbits living in dirty quarters and fed on contaminated pellets. The symptoms are weight loss, swollen belly, diarrhea and death. The liver is generally affected with lesions so that diagnosis during autopsy is easy. Control depends on sanitation. Prevention also depends on sanitation.

As I mentioned previously, cure is possible with drugs, but I think this is like sweeping dirt under the carpet. Surely you will learn that some rabbit feeds are furnished with a dose of sulfaquinoxaline to reduce the incidence of this disease. It is popular in big commercial establishments, but you would be better advised to practice better hygiene and nutrition with your small herd.

Another version of this disease affects the intestines rather than the liver, but as far as you are concerned this is the same problem and the same effort from you is required.

Rabbit Health

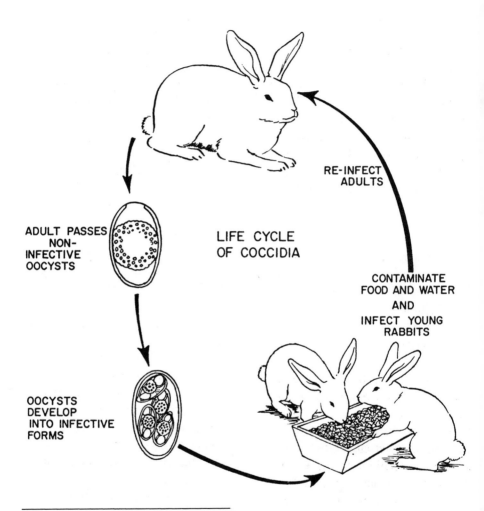

ADULT PASSES
NON-
INFECTIVE
OOCYSTS

LIFE CYCLE
OF COCCIDIA

RE-INFECT
ADULTS

CONTAMINATE
FOOD AND WATER
AND
INFECT YOUNG
RABBITS

OOCYSTS
DEVELOP
INTO INFECTIVE
FORMS

The life cycle of coccidia.

Rabbit Health

Molt

Molt is the natural process of hair replacement. It is generally accomplished in the late summer or autumn. Some animals get through a molt (sometimes spelled moult) in two or three weeks and others seem to take forever. A lot depends on genetics—some strains do it one way and some don't. This is a problem for exhibitors, but most other breeders don't pay much attention.

Fur buyers will notice the molt or lack of molt; but since a production rabbit is slaughtered as a juvenile and according to its weight, molt is not a consideration. No eight-week-old normal healthy rabbit is molting.

With very few exceptions, the pelt of a rabbit is a by-product and it accounts for only a small part of the total price, so the breeder enters the marketplace when the meat brings the best profit and the pelt goes along for the ride.

Stress

If your does don't produce and you cannot pin down the cause, consider the slim possibility that the animals are O.K. and that the fault lies with you. Too often, a doe resorbs or aborts or eats her young because she was frightened or stressed by a stranger or a loving and inquisitive child.

It is possible for a rabbit to break its own back by kicking out against a real or imagined enemy.

Malnourished animals are more vulnerable to stree-related problems.

Sudden or flashing lights upset rabbits.

Strange visitors and strange noises and strange odors in the vicinity of a kindling doe can wipe out a month of gestation in just a few minutes. Many breeders will not permit visitors near the breeders' hutches.

A newly introduced animal in a nearby cage, especially a buck, can cause stress on the others.

A sudden change in diet can cause stress. Once you establish a brand of pellets don't completely change it overnight. Don't suddenly pile fresh lettuce into a cage where it was previously unknown. Rabbits are creatures of habit.

Pain causes stress. If it seems that your technique with the tattoo causes pain, you should first chill the animal's ear for a few moments with a couple of ice cubes pressed gently against the area you plan to mark. This method causes a temporary numbing—almost like the

Rabbit Health

effect of novacaine in the dentist's office.

Some strains of certain varieties of any breed might be more vulnerable to stress than others. Some individuals may also be that way. Don't keep them.

Don't let all this frighten or stress *you*. Thousands of people breed millions of rabbits with little or no trouble. With thoughtful and humane care, you can too.

Steps to Help Prevent Disease in the Rabbitry

Disease is always an ominous threat when animals are kept in close confinement. The success of the rabbit man is largely determined by his ability to maintain a disease-free herd. To prevent disease outbreak take the following precautions:

ADULT TAPE WORM IN DOG'S INTESTINE

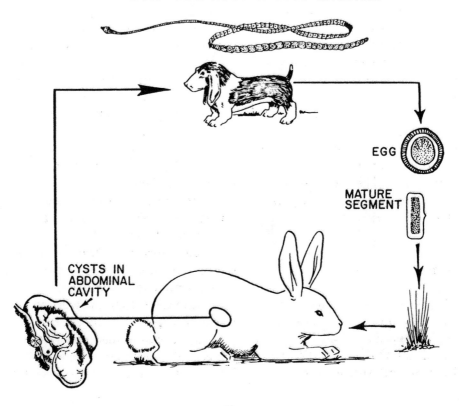

EGG

MATURE SEGMENT

CYSTS IN ABDOMINAL CAVITY

Rabbit Health

Handle your pet rabbit gently. If you frighten the little thing, it will show its fear by cowering. Learn how to hold your pet properly.

1. **Keep Feed and Water Clean** and fresh. Disease can be spread through water or feed.

2. **Clean and Disinfect the Nest Box** after each litter. Newborn rabbits are highly susceptible to disease and an infected nest box could be a real source of trouble.

3. **Learn to Recognize Rabbit Diseases.**

4. **Isolate all Rabbits** brought into rabbitry (show stock, new breeding stock, etc.) until positive that all danger of infecting the herd is over.

5. **Use Wire Floored Hutches** to keep the animal out of contact with their droppings and make an easier job of keeping the rabbitry sanitary and free of disease.

6. **Allow No Visitors** in the rabbitry, particularly pick-up men and other rabbit raisers who might unknowingly transmit disease to the herd.

7. **Sear Hutch Wire and Equipment** with a blow torch flame after marketing each litter. This will destroy most disease organisms and will protect the next occupant of the hutch.

8. **Select Disease-Resistant Animals** for breeding stock. They will pass on to their offspring greater ability to survive.

MOST COMMON RABBIT DISEASES

THEIR CAUSE AND CONTROL

NAME OF DISEASE	EXTERNAL SYMPTOMS	INTERNAL LESIONS	CAUSE	PREVENTION	TREATMENT
1. Coccidiosis	Listless, anemia, pot bellied, thin, loss of appetite, diarrhea.	White spotted liver, liver enlarged, inflamed intestine with occasional heavy mucous.	A protozoa—5 species of coccidia (Emira stiedea, etc.)	Use wire floored pens—Keep pens clean. Prevent fecal contamination of feed and water.	Feed Rapid Ade for 14 full days.
2. Ear mites mange	Scabs in ears, scratches head, may lose weight.	Scabs or crust in ear. Mites can be seen with microscope.	Ear mite.	Prevent contact with affected individuals.	Clean ear with cotton swab and apply weekly for 4 weeks 5% phenol in sweet oil or 5% chlordane solution.
3. Enteritis non-specific	Scours—dirty behind, feces dark, poorly formed, may stick to wire.	Inflamed intestines —feces in lower bowel soft, fecal pellet poorly formed.	Various types of bacteria or other cause of intestinal inflammation.	Prevent stress —wind, rain, poor housing. Do not troduce carrier animals.	
4. Eye infection Baby Adult	In baby rabbits, eyes may stick —inflamed eye— may have pus discharge.	Pus under lids in baby rabbits, inflamed eye and eyelids.	Various types of bacteria.	Prevent stress, ticularly cold drafts.	Apply specific eye ointment or antibiotic ointment daily.
5. Fungus infection	Scaly skin over shoulders or along back—hair thin— dandruff.	Excess dandruff and hair thin.	A non-specific fungus.	Prevent contact with affected animal.	Apply 2% Iysol solution to affected area every other day for 1 week.
6. Heat stroke	Panting—mouth open—quiet.	Muscle tissue appears par-boiled.	Excessive exposure to direct rays of sun. Lack of adequate ventilation.	Prevent direct exposure to sun or to poorly ventilated quarters.	Submerge in cold water—place in shaded area with adequate ventilation.
7. Ketosis	Occurs just before or just after kindling, listless, loss of appetite, diarrhea.	Liver large and light colored. Excessive fat in abdomen.	Overfatness—lack of exercise. Reduced feed intake, large litter.	Don't overfeed junior does—encourage exercise— provide palatable feed at kindling.	Prevent overfatness by limiting the feed to 4 to 6 oz. daily of Breeder Paks to Jr. does.
8. Mastitis— blue breast	Swollen milk gland—tender, may be dark colored—may abcess.	Caked mammary gland—surrounding tissues inflamed	Bacterial infection. Mechanical injury to mammary gland.	Prevent injury to mammary glands on edge of nest box. Clean and disinfect hutch.	Thoroughly disinfect nest box and hutch. Be sure nails or wire ends are not present on top edge of nest box.
9. Malocclusion or Buck Teeth	Difficulty in eating, wet about mouth—thin.	Lower teeth protrude—buck teeth —upper teeth long, curve into mouth.	Inherited defect— must be carried by both parents to show up.	Breed from malocclusion free parent stock.	Clip long teeth with sharp wire cutter.

MOST COMMON RABBIT DISEASES

THEIR CAUSE AND CONTROL

NAME OF DISEASE	EXTERNAL SYMPTOMS	INTERNAL LESIONS	CAUSE	PREVENTION	TREATMENT
10. *Muccoid Enteritis*	Most common at 5 to 8 weeks—excess thirst—pot bellied—diarrhea, feces watery or jelly-like.	Stomach full of water—intestines large—contain jelly like material.	Cause unknown.	Do not introduce carrier stock. Reduce stresses.	Do not introduce rabbits from affected herd. Keep hutches and nest boxes clean.
11. *Pasteurellosis— Hemorrhagic septicemia*	Most common in fryers, listless, pot bellied, diarrhea, rapid breathing.	Pneumonia with abcesses in lungs —in does, may have nasal discharge or excess liquid in abdominal cavity in fryers.	Specific bacterial infection, coupled with stress factors.	Don't introduce carrier stock— avoid stress where possible.	Inject with 25,000 to 50,000 units penicillin and 100 units dehydro-streptomycin.
12. *Pneumonia*	Most common in does, rapid breathing, nasal discharge, head held high.	Abcesses in lungs or solidified areas in lungs.	Bacterial or virus infection, associated with stress factors.	Avoid stress and use good sanitation.	Inject as with pasteurellosis.
13. *Ringworm*	Loss of hair around, face, ears, over body; skin inflamed in rings. Do not confuse with hair pulling.	Circular areas of loss of hair, scaly with red inflamed skin.	A specific fungus.	Do not introduce contaminated animal.	Disinfect hutch, then dip all affected rabbits in lime sulphur dip.
14. *Rickets*	Occurs only in dark rabbitry. Fore or hind legs crooked. Spraddle legged.	Front or rear legs crooked. Ribs beaded, bones fragile.	Calcium, phosphorus or vitamin D deficiency. No access to sunlight.	Feed a balanced ration and supply direct sunshine.	Supply ample calcium, phosphorus and vitamin D as supplied in Albers Rabbit feed.
15. *Sore hocks*	Hunched up or lies stretched out. Pain on walking. Thin.	Scabs usually on bottom of hind feet, may be on front also.	Injury followed by infection of sole of foot. Injury from floor with infections. Insufficient floor support.	Select for thick foot pads. Avoid sharp protruding objects, wire or nails in floor. Avoid wet litter or manure accumulation in pen.	Destroy seriously affected animals. Mildly affected, clean foot with soap and water. Apply saturated solution of Bluestone once weekly. Supply a soft bedding.
16. *Slobbers*	Wet about face. Do not confuse with malocclusion. Face may show swelling.	Swollen cheek may contain pus. Intestines lack tone, liquid content.	Excessive feeding of green feed. Indigestion. Abscessed molar teeth.	Do not feed excessive greens.	Cut back on green feed. Extract abcessed tooth.
17. *Vent disease*	Blisters or dark scabs on external genitals, parts swollen.	Usually no internal lesions.	Urine burn, hot metal floor burn or infection by a spirochete.	Do not allow manure or moisture to accumulate in pens. Do not introduce affected or carrier animals.	Clean off scabs and apply antibiotic ointment or powder.

Harmful and Safe Plants

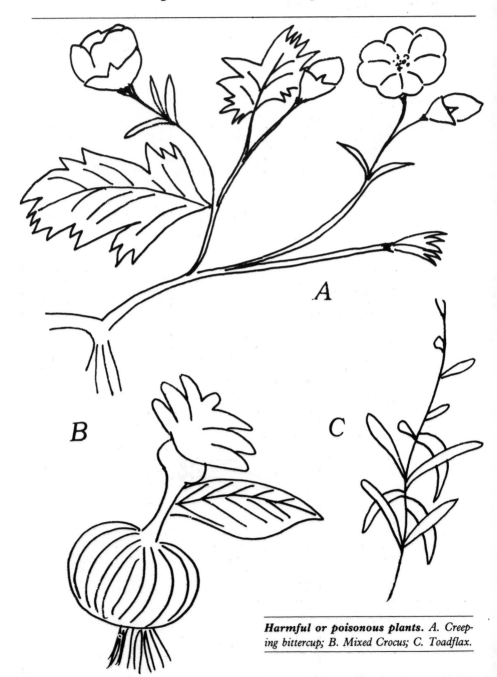

Harmful or poisonous plants. A. Creeping bittercup; B. Mixed Crocus; C. Toadflax.

Harmful and Safe Plants

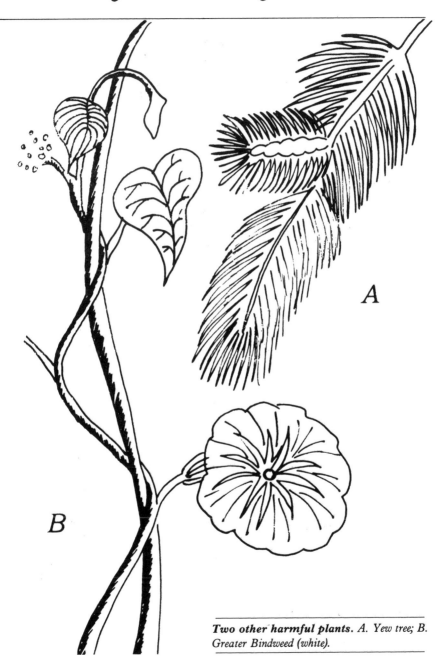

A

B

Two other harmful plants. *A. Yew tree; B.*
Greater Bindweed (white).

Harmful and Safe Plants

Safe plants for rabbits. A. Yellow Ground-sel; B. Raspberry; C. Coltswood (it's yellow); D. Shepherd's purse; E. White Ground Elder.

More to Read

Here is a short list of things for you to read which are more detailed or cover aspects which I glossed over in this introductory book.

Commercial Rabbit Raising, Agriculture Handbook No. 309 by the U.S. Department of Agriculture, is available from the Superintendent of Documents, Washington DC 20402 Catalog No. A 1.76:309/2. This booklet gets into slaughtering, rabbit skins, Angora wool and economics. It is clear and authoritative but weak on descriptions of the breeds.

The Domestic Rabbit by J.C. Sandford is a British export. Now in its third edition, it is published by Crosby Lockwood Staples and Granada Publishing. This 258-page book by a recognized authority has a wealth of information and some good black and white photographs. It is most comprehensive.

Encyclopedia of Pet Rabbits is by D. Robinson. Published by T.F.H. Publications, Inc., this is by far the best book for color photographs of rabbit breeds and varieties. There are 100 pages of color. The text is clear and thorough. It is a complete book for a pet keeper.

Raising Rabbits is a really good soft-cover book that is in agreement with its title. It is by Ann Kanable and is published by Rodale Press. Simple drawings and a few black and white photographs are clear and easy to understand. Ann Kanable's rabbits are raised economically, simply and humanely; and they are slaughtered, skinned and dressed in methods anyone can accomplish inexpensively. This is a most practical book for someone who wants to put meat on the table.

Watership Down is by Richard Adams. When you think you know all you need to know about rabbits (but you still have a few tiny lingering doubts), you might enjoy reading this natural history of the wild version of this species. It is in the form of a novel, but the facts about these animals are based on scientific evidence. No rabbit keeper should get through life without having read it.

Glossary

THE T.F.H. BOOK OF PET RABBITS
By Bob Bennett
ISBN 0-87666-815-5
T.F.H. HP-014

Contents: Introduction, The Rabbit as a Pet. Choosing a Pet Rabbit. Feeding. Housing and Equipment for Your Pet. Handling and Health Care. Rabbit Shows. Mating and Raising Rabbits.
Audience: This book, written by the author of the Boy Scout manual about rabbit raising, is full of down-to-earth advice and rabbit lore. For those thinking of purchasing a pet rabbit, there is no better place to begin than the *T.F.H. Book of Pet Rabbits*, which will start the new owner out *right!* Ages 14 and up.
Hard cover, 8½ x 11", 80 pages
85 full-color photos

STARTING RIGHT WITH RABBITS
By Mervin F. Roberts
ISBN 0-87666-814-7
TFH PS-796

Contents: Introduction. Rabbit Cousins. Rabbits in Legend and Literature. What Rabbits Eat. Choosing a Pet Rabbit. Handling Rabbits. Diseases. Breeding. Genetics. Economics. Odd Facts about Rabbits.
Hard cover, 5½ x 8", 128 pages
78 full-color photos, 10 black and white photos

ALL ABOUT RABBITS
By Howard Hirschhorn
ISBN 0-87666-760-4
T.F.H. M-543
Contents: What About Rabbits? Rabbit Housing. Shipment Of Rabbits. Feeding. Health. Breeding. Rabbit Handling. Rabbit Clubs. Standards And Shows. Profitable Commerical Aspects Of Raising Animals. Final Thoughts On Rabbit Theory.
Hard cover, 5½ x 8", 96 pages
32 color photos; 39 black and white photos.

Angora—A rabbit whose hair is about or longer than three inches. The name probably derives from Ankara, Turkey, where the breed may have originated.

Breed—A race of rabbits that produces similar offspring. Not a mongrel. Americans recognize over 25 breeds.

Breeder—A person who breeds rabbits. Also a rabbit which is used for breeding.

Buck—A male rabbit.

Bunny—*See* **Kitten.**

Cage—Indoor rabbit housing. A screened enclosure which might not have a box or covered part. *See* **Hutch.**

Charlie—A color pattern that is pale, washed out.

Cobby—Stocky, short. An English Lop is cobby.

Density—Thickness of the fur, a relative term.

Doe—A female rabbit.

Dwarf—A breed of rabbit which when mature weighs not more than three pounds.

False pregnancy—When a female rabbit fools you.

Flyback—Fur that quickly returns to normal after it has been blown or brushed the wrong way.

Gestation—The time between

Glossary

mating and kindling, usually 31 days.

Giant—A rabbit which when mature weighs twelve pounds or more.

Guard hair—Coarse long hairs, normal on most breeds.

Hutch—Outdoor rabbit housing consisting of a screened part and a covered part. *See* **Cage.**

Junior—A rabbit less than six months old.

Kindling—Giving birth.

Kitten—An unweaned rabbit. Also bunny, pup.

Line—A strain of rabbits.

Line-breeding—Deliberate inbreeding to retain the features of a strain. A buck might be mated to his daughters and to his granddaughters.

Mating—Copulation. The fertilizing of a female rabbit.

Medium—A rabbit which when mature weighs more than three but less than twelve pounds.

Molt—Shedding of fur. A normal annual affair. Also spelled moult.

Pair—Rabbits that have been mated.

Palpation—An examination of a doe by an expert to determine if she is pregnant. A beginner should beware—more harm than good can come of this if it is done improperly.

Pup—*See* **Kitten.**

Purebred—A rabbit of a recognized breed.

Racy—Slender, slim. A Himalayan is racy; a Belgian Hare is even more so.

Rex—Hair like plush velvet. A short-haired rabbit.

Satin—Hair short, transparent and lustrous.

Senior—A fully mature rabbit. A large breed is a senior at eight months. A small breed is a senior at six months.

Strain—A family of rabbits that exhibit the same genetic characteristics.

Test-mating—The doe is returned to the buck eight days after mating. If she rejects him, she is probably pregnant.

Ticking—Contrast between color of guard hairs and that of under-fur.

Type—Conformation as racy or cobby.

Under-fur—Short, soft hairs normal in most breeds.

Under-wool—Under-fur.

Variety—A color or size designation within a breed of rabbits. Americans recognize about 75 varieties distributed among over 25 breeds.

Chart of Breeds

Breed	Standard Mature Weight		Registration Weight (for entering in shows)		Primary Utility Value
	Buck	Doe	Buck	Doe	
American (Blue and White)	9	10	8-10	9-11	Meat, fur, show
American Silver Fox (Black and Blue)	9½	10½	8-11	9-12	Show
Angora Woolers.	6 or over	6 or over	5½-7½	5½-8	Wool and show
Belgian Hare.	8	8	6-8	6-9	Show
Beveren, Blue	9½	10½	7-10	9-11	Show
Beveren, White	9	10	8-10	9-11	Show
Blue Vienna	9-10	10-11	8-11	9-12	Fur and show
Californians	9	9½	8-10	8½-10½	Meat, show, fur
Castorrex	8	9	7	8 or over	Fur and show
Champagne de Argent	10	10½	9-11	9½-12	Fur and show
Checker Giant	11	12	11	12 or over	Meat and show
Chinchilla, American . . . . . Chinchilla Giant	13-14	14-15	12-15	13-16	Fur, meat, show
Chinchilla, American Heavyweight	10	11	9-11	10-12	Fur, meat, show
Chinchilla, American Standard.	6½	7	6-7½	6¼-8	Fur, meat, show
Chinchillarex	8	9	7 or over	8 or over	Fur and show
Dutch, Black, Blue, Grey, Steel Grey, Tortoise and A. O. C. . . .	4½	4½	3½-5½	3½-5½	Show
English (spots any color)	6-8	6-8	5-8	5-8	Show
Flemish Giants, Steel Grey, Light Grey, Black, White, Blue	12 or over	15 or over	12 or over	13 or over	Meat, show, fur
Flemish Giant, Sandy Grey	14 or over	16 or over	12 or over	13 or over	Meat, show, fur
Havana, Standard	6	6	5-7	5-7	Fur and show
Havanarex	8	9	7 or over	8 or over	Fur and show
Himalayan	3½	3½	2½-5	2½-5	Show and fur
Lilac	6-7	6½-7½	5½-8½	6-9	Fur and show
Lops (English) — earage not less than 16 in. from tip to tip	10	11	Not under 9	Not under 10	Show
Lops (French) — earage not less than 14 in. from tip to tip	10	11	Not under 9	Not under 10	Show
New Zealand (Red and White) seniors only	10	11	9-11	10-12	Meat, show, fur
Polish	2½	2½	Not over 3½	Not over 3½	Show and fur
Rex	7	8	7	8	Show and fur
Sable, American	8	9	Not under 7	Not under 8	Fur and show
Satin	8½	9½	8½	9½	Show and fur
Sablerex	8	9	Not under 7	Not under 8	Fur and show
Silver (Grey, Fawn, Brown) . . .	6	6	4-7	4-7	Show and fur
Silver Marten (Black and Blue)	7½	8½	6½-8½	7½-9½	Show and fur
Tans (Black and Blue)	4-5	4½-5½	4-5½	4-6	Show